CLASSIC SPORTSCARS

CLASSIC SPORTSCARS

Paul Badré and Alberto Martinez

Exeter Books
NEW YORK
A Bison Book

Originally published as *Voitures de Sport*, © 1987 E.P.A.

First published in USA 1988
by Exeter Books
Distributed by Bookthrift
Exeter is a trademark of Bookthrift Marketing, Inc.
Bookthrift is a registered trademark of Bookthrift Marketing
New York, New York

ISBN 0-671-08962-5

Printed in Hong Kong

Design: Yves Le Ray
All photographs by Alberto Martinez except for pages 42 and 43 (Xavier de Nombel) and 88, 87 and 125 (William Borel).

CONTENTS

AC COBRA

AC Cobra 289

The traditional open sportscar in which the driver sat exposed to the elements, died out at the end of the sixties. It was killed off by the 'Grand Tourisme' or GT coupe in which the driver sat snugly in a protective shell. This demise was hardly surprising – the trouble with being on top is that there is only one way to go – and that's down. The 1963 AC Cobra was the very epitome of the open sportscar. But like all thoroughbreds the Cobra was something of a hybrid. In 1953, AC had launched the Ace, with its chassis designed by John Tojeiro and a body that was a carbon copy of the boat-shaped racing shell that Ferrari was using at the time. Until 1961, the Ace had led a rather quiet life. At first, it had been powered by a 6-cylinder AC engine designed in 1919. Later on it was given a Bristol 2-liter which was not much more up-to-date – a 6-cylinder BMW engine of 1933-37 vintage. But all this changed when an enterprising individual called Carroll Shelby came along and proceeded to completely transform the rather sedate Ace.

Shelby was a Texan who had been racing for years in Europe (mainly for Aston Martin) and had gone to settle in California. In 1961 AC found itself without an engine, Bristol having switched to the Chrysler V8. Shelby presented AC with an ideal, readymade solution: the new smallblock Ford V8. The Ford wasn't any heavier or more cumbersome than the old, heavily souped-up Bristol and could, without overexerting itself, turn out twice the horsepower. So, from 1962 onward the AC left the factory in Britain without an engine and headed west to the Californian workshop of engine-whiz Dr Shelby. He proceeded to pack into each Cobra more power and more sex appeal than anyone had dreamed of.

In its deadliest, most lethal form the Cobra 289 packed under its hood 330 horsepower. The 4.73-liter V8 Ford had at its disposal 4 dual-throat Weber carburetors, a special camshaft and larger diameter valves. Its top speed was more than 150mph and it could cover the standing quarter mile in a shade over 14 seconds. At first, it was named the 'Shelby AC Cobra' but soon it was simply known as the Cobra, the name which Carroll Shelby had long before chosen for his future sportscar. The phrase 'powered by Ford' was added at the insistence of Ford's head office in Dearborn. The company was, after all, providing technical assistance as well as the engines.

The cockpit of the Cobra 289 was stripped to bare essentials. By cutting back on accessories the makers were able to achieve the sort of power-to-weight ratio more usually associated with big superbikes. The V8 Ford engine developed at least twice the power of the original Bristol and was no heavier. The gearbox was also provided by Ford. Between 1962 and 1967, 630 Cobras were built, including 60 units powered by the 4.26-liter V8. The AC steering system was soon replaced by rack-and-pinion steering copied from the MGB, and the English Smith instrumentation gave way, at Ford's insistence, to a set of dials by Stewart-Warner.

ASTON MARTIN

Aston Martin DB2/4

The modern GT arrived on the scene in England and Italy at roughly the same time, in 1950. But while the Lancia Aurelia Gran Turismo was the result of a long maturing process, the Aston Martin DB2 was more like a hastily improvised cocktail, developed in virtually no time at all – not that this is necessarily a bad thing.

It was in 1947 that industrialist David Brown bought Aston Martin. He was the fifth owner of the firm which had been set up originally in about 1914. It was a self-indulgent present to himself, intended more to satisfy his passion for motor racing than in the hopes of making money. He was right on both counts – Aston Martin didn't earn him a cent, but throughout the 1950s it was to perform splendidly in competition and would even make its mark on the Formula 1 scene.

In 1949, the 24-Hour Race was about to be run for the first time since the war. Aston Martin had entered every race since 1928. Then fate reared its head. The only car Aston Martin had available was a 2-liter machine which concealed under its generous bodywork a rather interesting rectangular-section tube-chassis, with coil springs on all corners linked to twin trailing arms. At the rear was a beam axle, while the front wheels were independent. Having gotten that far, Aston's engineer Claude Hill ran out of ideas – the engine was an ordinary ohc 4-cylinder unit. The body-styling with its overblown wings didn't do much for the car's appeal either, and up front the lack of taste was consummated by a high neoclassical radiator grille flanked by two side vents.

Aston Martin prepared no less than six cars for Le Mans. Three of these were to be streamlined coupes rather like small Italian saloons with a touch of British stiffness. The radiator, lowered like the rest, was a foretaste of the classical look that the firm would always adhere to from then on. One of these coupes was a little different. In a last-minute decision, David Brown gave it a 2.6-liter, 6-cylinder engine which had been built by another firm he had bought – Lagonda.

This dual overhead camshaft engine had been designed by W O Bentley, who had already begun to suspect that the V12 he had built before the war would have no place in the austere years that followed that conflict. Developing 107hp at 5000rpm, it was unusual in having a crankshaft gripped in light alloy bushes sticking through one end of the crankcase. Lagonda had unfortunately stuck the engine into a body which was as ugly as it was old-fashioned.

The Aston Martin DB2/4 was unwittingly the first high-performance hatchback. This idea was taken up later by Jaguar and then by Aston Martin again with several DB5s. The original 1950 DB2 had only two seats and no trunk. Aston Martin converted it into a 2 + 2 by incorporating a third door, a folding rear seat and lots of ingenuity. The classical Aston Martin radiator grille (photo opposite) was originally a three-piece design, but the three came together after the car suffered overheating problems in the 1949 Le Mans race. The design still bears traces of the original today. The dashboard of the DB2/4 is a direct descendant of those on the prewar Lagondas. The traditional look of the dashboard provides a marked contrast to the modern look of the bodystyling.

To race at Le Mans, the Aston's power was increased to 137hp and the light coupe could then touch 125mph. But the car didn't get very far – the engine had been installed with so much haste that its cooling system was unsatisfactory. The coupe dropped out before the end of the first hour, but this no longer seemed to matter. Aston Martin had come up with a winning combination – a streamlined body, a modern chassis and a thoroughbred engine. The coupe went into the Aston Martin catalog in April 1950 as the DB2 (David Brown Mark II) and thus the original 2-liter car became the DB1.

The DB2 was offered in two versions – the ordinary model (107hp like the Lagonda) and the 116hp Vantage. The much sought-after Vantage could do better than 110mph. Light, stable and easy to handle, the Vantage was a definite success and by the following year the Aston faithful (by now, a new successor had come along) were demanding a version with more than two seats. Trying to make what came to be known as a 2 + 2, the firm found a very simple solution and in fact ended up with a model which was later known as a hatchback. The spare wheel went underneath and a forward-folding rear seat was installed in the rear section. With typical British humor, this car was christened the DB2/4 and was launched in 1953. The following year its engine capacity was increased to 3 liters in order to offset the increase in weight.

Pretty soon after this Aston Martin found that the 2 + 2 version could do with a number of changes; the DB4 was the next model to emerge from the drawing board.

Aston Martin DB5

There has been a long line of Aston Martins since the war, the same car re-emerging each time with little changed. Somewhere between the sleek, youthful DB2 and the amply rounded DBS V8 came the rather mature DB5. The DB5 is a serious car, meant for the top executive in a hurry on those days when the Beechcraft is in for its checkup. The high performance and the sporty handling are not intended as ends in themselves but simply as a means of ensuring the owner will arrive at the next board meeting feeling fresh and relaxed.

The DB5 replaced the DB4 in 1964, but there were only minor differences – engine capacity increased to 4 liters, a 5-speed ZF gearbox and a restyled front with streamlined headlamps in the same vein as the DB4GT. There was really not much justification for calling it a new model, but Aston Martin undoubtedly wanted customers to forget the teething troubles of the original model. The DB5 was the DB4 come of age.

Aston Martin had started thinking about a 4-seater as far back as 1955, in response to criticism of the DB2. The project was given to John Wyer, who had become technical director after spending five years as chief of testing. The new car's engine was designed by Tadek Marek, who looked to the race-tested W O Bentley engine for inspiration. It was still a straight six with dual overhead chain-driven camshaft and cylinder-liners of chrome-vanadium steel, but this time the 3.67-liter (92mm × 92mm) powerplant was built entirely of light alloy with a standard crankcase. It could develop 263hp at 5500rpm with its twin SU carburetors. A special dual-ignition cylinder head and three dual-throat Webers could boost its power to 344hp.

The chassis designed by Harold Beach took up the DB2/4 fixed rear axle with double trailing arms but had in addition a Watt steering and braking system. Up front, the neo-Porsche suspension gave way to a standard wishbone set-up. The Aston Martin body-builder, Tickford, dressed the DB4 using the Superleggera technique under license from Carrozzeria Touring – light alloy on a tubular-steel skeleton. The bodystyling also bore the Touring stamp. The new car was to have had the DB2/4's tubular frame but Touring insisted on a platform chassis.

The DB4 was launched in October 1958, just in time for the opening of Britain's first freeway. But the new Aston Martin had the annoying habit of losing oil pressure when it was hot, through excessive expansion of its light alloy cylinder block. This and a few other faults were progressively ironed out, and Aston Martin's years of patient work were rewarded when the DB5 was finally released in 1963. The DB5 had an increased engine capacity of 4 liters (92mm × 96mm) and could turn out 286hp. The car was a little heavier than the DB4 (3300 pounds as against 2850 pounds) but its performance remained unchanged – it boasted a top speed of nearly 150mph. And this time the Aston Martin was very definitely solid and stable.

As someone once remarked rather wittily, when Aston Martin released their new model, 'The DB5 is the ideal second car for someone who already owns a Mini.'

The DB5 has a dashboard that is rather businesslike and perhaps a little too sedate. But then the DB5 was not a car meant for kids. The top executive making a lightning dash from one city to another would have felt quite at home behind the wheel of his Aston Martin. The driver's seat of the DB5 had a somber feel about it – comfort and convenience abounded but there was not the slightest touch of fantasy. Instead of titivating the driver's seat Aston Martin had chosen to offer the driver of a DB5 more useful additions, like the 5-speed ZF gearbox or the optional Borg-Warner automatic. And the Rudge wire wheel was having its last fling with the DB5 before the advent of light alloy rims.

1782 QJ95

DB
5

Aston Martin Volante

Aston Martin, the car manufacturing company which burst on the scene in the fifties and remained at the forefront of British motoring for two decades, underwent a dramatic transformation in the seventies. The company stopped hunting in Jaguar territory and set its sights on Rolls-Royce terrain instead.

The man responsible for this development was a well-connected millionaire called Peter Sprague. It was Sprague who, in 1975, saved the firm from landing up in the hands of the receivers. This was the sixth or seventh change of ownership Aston Martin had undergone, but this time was different – for the first time in its history the firm was about to start making money. Sprague managed this by simply doubling the price of his cars. It was a much more astute move than it seemed. Sprague knew what he was doing – like it or not, a car buyer, like any other consumer, can be favorably impressed by a high price tag. So when the price rocketed, so did Aston Martin's prestige. Like the Rolls-Royce, the Aston Martin was soon to join the ranks of those objects whose primary merit is that they cost a great deal. The upshot of all this was that by the late eighties the Volante convertible cost $140,000. The total production run in 1985 was less than 200 cars. And the product has undergone only minor modifications in recent years.

However, today's Aston Martins are not content to be exceptional only because of their price tag. The quality of their workmanship is probably unequaled and certainly unsurpassed anywhere. The metal-workers certainly deserve a good reputation. When you run your hand over the body you can only just tell where the door meets the body panel. The factory takes a little more than four months to produce each car. The bodywork alone is given about 20 separate coats of paint. And the finishing coats are not applied until after the roadtesting (averaging 120 miles) has been completed.

The assembly of each engine is entrusted to a single technician who 'signs' his work by affixing a metal nameplate onto the rocker cover. Aston

The graceful lines of the Volante have a sculptural feel about them. Aston Martin bodywork has a well-deserved reputation for perfection. This is particularly evident in the streamlined doors.

Martin wants to make it clear to each customer that their vehicles have not been built by robots but by craftsmen who possess a love for their trade. The finished car then spends five hours on the test bench. So each car is delivered to its new owner in a condition comparable to that of the new clothes that Beau Brummel was reputed to have had 'run in' for a few days by his personal valet before he finally wore them himself. If you have the impression that the Aston Martin factory is a little enclave where time has stood still, you're probably not far from the truth.

The inspiration for the current models is the DBS which was unveiled at the London Motor Show back in 1966. That model had been designed around a V8 engine which was not ready in time, so the DBS had to be content with the 6-cylinder unit from the old DB6. The body-styling was the work of designer William Towns. It was he who was later to be responsible for the outrageous (the word is too kind . . .) body that went onto the Aston Martin Lagonda. The V8 made its competition debut powering the Lola coupes for the 1967 racing season. Its perform-

The driving compartment of the Volante has been called a symphony in leather and polished walnut. It is in details like these that the transformation of Aston Martins into cars intended for the luxury market becomes most apparent.

ance was quite catastrophic and several years were to go by before Aston Martin was able to offer the V8 to the public. And all this time, the DBS was doing very little to promote Aston Martin's name. Much too heavy for its 6-cylinder engine, it turned out to be a bit of a tank.

It was only in 1970 that the DBS was given the V8 that it was designed for, along with a few quite positive styling improvements added by William Towns. The DBS was at that stage by and large the same car as the one being built today. Among the few things to have changed is the Bosch fuel injection which Aston Martin never quite mastered. In 1972 this gave way to a set of four dual-throat Weber carburetors, so doing away with the last anachronism in what is, at heart, really a 'vintage' car. After that, the V8 probably developed close to 250hp at 5000rpm. But that is pure speculation because Aston Martin chose the launching of its V8 to adopt a policy of no longer divulging the horsepower of its engines. A more powerful version (300hp?) powered the V8 Vantage unveiled in 1977.

The Volante convertible went on sale in 1978. The new model was priced at $66,000 and came complete with air-conditioning and a stereo radio. The Volante proved such a success that by the following year the normal coupe was no longer produced except for special orders. In 1979 the air scoop on the hood was eliminated in order to conform to American standards.

Today's 135mph thoroughbred has a V8 engine with a capacity of 5340cc (100mm × 85mm). It has dual overhead chain-driven cams on each bank of cylinders. The gearbox is either a 5-speed ZF manual or a Chrysler 3-speed automatic. The bodywork is light alloy mounted on a steel platform chassis. The rack-and-pinion steering is power-assisted, which comes in handy in a car whose weight is close on two tons.

The instrument panel of the Volante has a classical elegance. The luxurious materials used and the high quality of the craftsmanship complement the simplicity of the design. Despite an all-up weight approaching two tons, the Aston Martin can top the 135mph mark and still maintain a companionable silence. The car shown here was built in 1983, but any confusion with an earlier or later vintage is quite forgivable; the confusion is, if anything, actively encouraged by the makers.

With its wide track of 59 inches the Volante gives a reassuring impression of stability which is certainly not belied by its performance. The Volante is today one of the last bastions of a tradition of quality which is 'still there once you've forgotten how much you paid.'

BMW

BMW M1

The BMW M1 had everything going for it. The combination of German know-how and Italian creative genius should have produced an earth-shaker of a car. But something went wrong. Instead of sound and fury, all the M1 could manage in the end was a whimper. The car itself was certainly no misfit. All that was lacking for it to be a resounding success was that little dash of extravagance that even the Porsche 911 Turbo managed to possess.

The original idea was to design a small racing saloon along the lines of the Lancia Stratos and produce the 400 cars necessary for a Group 4 rating. The project was launched around 1975 by Jochen Neerspach, director of the BMW offshoot Motorsport, which is in effect its racing division. The project was too big for Motorsport to handle alone but not quite meaty enough for the head office, so Neerspach approached Lamborghini, who were hardly snowed under with work. In the end the Italian company handled the overall conception of the car. It turned out to be a 2-seater coupe with a mid-engine, a tubular frame and a fiberglass body. The Motorsport powerplant comprised the 3.5-liter, 6-cylinder engine from the 646 CSi topped with a dual camshaft, 24-valve head.

By mid-1977, the prototype had begun road-tests and the production line was being set up in the Lamborghini factory, but financially the Italian firm had its back to the wall. By the following year, the delays were so serious that BMW canceled the contract. Production of the M1 ('M' for Motorsport) was then entrusted to the Stuttgart body-builder Baur. The BMW staff did not, however, receive the foreign-looking car which they had just inherited with much enthusiasm.

BMW had called on Giorgio Giugiaro to design a body. The worthy Italian, deluged with orders, devoted no imagination at all to the M1 and it ended up looking like a Lotus Esprit with a vaguely BMW radiator grille. The austerity of the equipment that the German workers bestowed on their creation did little to enhance its appeal. And with the cars costing more than $40,000 each, BMW was not besieged by buyers. The M1 failed to enjoy much success on the racing circuit too, thanks to the lackluster support it received from the factory. The BMW M1 had finally been betrayed by all those who should have been able to guarantee its success.

The BMW M1 was an international mega-production with a very impressive list of credits – it had an engine by BMW, styling by Giugiaro, and it was built by Lamborghini. By rights, the M1 should have been a roaring success. Its 3.5-liter engine with twin overhead camshafts and four valves per cylinder developed 277hp at 6500rpm with that mellow smoothness that 6-cylinder BMW engines have become famous for. It had a top speed of 160mph coupled with absolutely faultless road-handling. But one by one all those who contributed to the M1's development let it down in the end. Lamborghini could not keep to their side of the bargain, BMW took over the project without much conviction and Giugiaro made no more than a half-hearted attempt at bodystyling for the car. In the end, 450 of the cars were built. But the M1 arrived far too late and its price was far too high. Its career was rather ignominious, both on the racing circuit and on the showroom floor. Yet the first mid-engined BMW deserved a far better fate.

CAMPAGNOLO

BMW M1

CHEVROLET
CORVETTE

CHEVROLET

Chevrolet Corvette 1958

It seems there are still some top executives at General Motors who haven't even heard of the Corvette. Perhaps they can be forgiven for never having noticed it – sales of the GM sportscar have never even reached one percent of the corporation's overall figure.

On the other hand, in the world of car lovers, the Corvette has more fans than any other car GM has ever built. It has its own cult following and its own mystique, with secret passwords known only to the initiates. One of the reasons for this is that for the American motoring scene the Corvette has always been something of a loner. It has always kept far from the madding crowd of 'other' cars, evolving in its own particular way without sacrificing its individuality.

In the beginning, the Corvette's personality was slow to crystallize. The man responsible for creating the Corvette was Harley Earl, who was the styling chief at General Motors for 30 years. It was Earl who gave the American automobile its distinctive touches: the tail fins, the bullet-like bumper guards, the panoramic windshield. At the start of the fifties, Earl's sons were about to go to college. Suddenly he realized that there was no American car built for the youth market and that young Americans were turning to imported European sportscars instead. Earl was not a man to let an idea gather dust and immediately mapped out for his associates a program for the future Corvette – 'A simple, cheap, European-style convertible sportscar.'

The first of the Corvettes was unveiled at the 1953 Motorama, the prestigious showplace where GM (in other words, Harley Earl) liked to present its latest creations. The Corvette was a strange-looking machine, a sort of remake of the Ferrari racing shell as seen through the eyes of a Texan. The body was made of fiberglass for the simple reason that GM found that material easier to work with on its prototypes. The basic shape was simple and fairly pleasing, but Earl had deemed it necessary to add his latest find – the

In 1958 the Corvette finally came close to the original sales target of 10.000 cars a year that had been set for it. By then, the headlamps had been doubled up and the hubcaps more than ever resembled fancy cake-molds. Two decorative trim-strips had appeared on the trunk lid. This is a non-standard hood with no blisters but a row of louvers. As the advertising blurb never tired of mentioning, the 283ci Chevrolet V8 with its Rochester fuel injection could turn out 283hp. The 1958 version was a little heavier and its performance was not quite as stunning as that of its predecessor. But for the men at head office it was an absolute winner – sales of the 1958 Corvette climbed to 9168 units after the sluggish 6340 recorded the previous year.

panoramic windshield. He also gave his creation protruding taillamps that resembled bombs complete with their fins. The streamlined headlamps were positioned behind stone guards and the doors had neither wind-down windows nor outside handles. These 'sporty' features did not quite fit in with the rather tame 6-cylinder Chevrolet engine, especially when it was coupled to a lethargic Powerglide automatic transmission.

Despite its shortcomings, the Corvette was marketed in this form during 1954 and 1955. But at $3500 (almost twice as much as a Cadillac convertible) it was not a huge success. 3600 were sold in 1954 and only 700 the following year, far below the sales target of 10,000 per year that had been set. By 1956 the bombs at the tail had gone and the headlamps had popped up again. The front wheel arch had been flared backward and the doors finally had wind-down windows. But above all, the Corvette now had a Chevrolet V8 developed by Ed Cole. With its twin carburetors, the new engine developed 225hp and the Corvette at last took off. The 1957 model was given a Rochester fuel-injection system that boosted the power to 283hp and pushed the top speed to 125mph. By 1958 the car had grown ten inches in length, two inches in width and 200 pounds in weight. A synchronized 4-speed Borg-Warner gearbox completed the transformation to the 1958 Corvette shown here.

ZG · 14032

GOODYEAR
EAGLE GT

Chevrolet Corvette Stingray

The Corvette only really started to take off when Zora Arkus-Duntov took charge of its development. This cosmopolitan Belgian had grown up in Russia and studied in Germany before embarking on a career in the United States, where he had developed the Ardun cylinder head for the V8 Ford. In 1953 he moved to Chevrolet, determined to make the Corvette his 'baby.' His influence was first felt in the 1957 model for which he designed a special camshaft. He improved the roadholding of the 1958 model quite spectacularly by a simple rearrangement of the suspension. But his masterpiece was the famous Sting Ray released at the end of 1962.

All that the new Sting Ray retained of the previous model was the 5.36-liter V8 with its Rochester fuel injection. Everything else had gone back to the drawing board, starting with the 98-inch wheelbase chassis. Up front there was a standard Chevrolet suspension but the rear wheels had independent suspension.

Yet these novelties were quite eclipsed by the sensational body designed for the Sting Ray by Bill Mitchell, Harley Earl's successor as GM style chief. One of the main features had been taken from the very first Sting Ray, which was a racing shell derived from the original Corvette. This was a sort of one-block body that reduced the front and rear quarter panels to simple shaped humps above each wheel. The headlamps were fully retractable.

The 1963 Sting Ray came in two different versions, convertible or coupe. This was the first of the Corvette coupes. It had a distinctive rear-pointed cockpit canopy with two-piece rear window which immediately earned it pride of place among the Corvette cult following. The Sting Ray was a commercial success too – after its introduction, Corvette could boast sales which regularly topped 20,000 per year. Further developments continued to take place.

The following year, the split rear window, more spectacular than practical, disappeared from the coupe. In 1965, the Sting Ray was given disc brakes on all four wheels and a 6.5-liter V8 that developed 425hp. In 1966 the engine capacity

With its clean, flowing muscular lines the Stingray has lots of class. One of its striking features is the way the body slims down like a Coke bottle near the driver's seat. This effect is accentuated by the styling of the stainless-steel rocker panel. The cabin roof is made up of two removable sections which leave a tee-shaped roll-bar. The Stingray retains two features of the original 1953 Corvette – the bodywork is still made of fiberglass and there is still no trunk. The 1969 model shown here is equipped with the more powerful of the 'production' engines, the 7-liter bigblock V8. This engine was capable of turning out 435hp at 4800rpm with all the smoothness of a rocket leaving the launchpad. The Stingray's top speed was around 155mph.

was raised to 7 liters but the Rochester fuel injection was replaced by Holley Carburetors. The 1967 models were given Rochester carburetors which were not quite as thirsty as the Holleys.

For its 1968 release the Corvette was given a new look but the mechanics remained for the most part unchanged. The new bodystyling was inspired by the experimental Mako Shark II. This was the work of designer David Holls, who had also worked on the Oprel GT. The 1968 Corvette had a slimmed-down mid-section, rather like a Coke bottle. This styling effect highlighted both the muscled bulge of the wheels and the long slender hood. On the whole, the new bodystyling remained faithful to the spirit of the original Sting Ray. The 1968 model had a number of other new features too – vacuum-controlled flip-up headlamps, a cover for the windshield wipers, plus a full fiber-optic monitoring system.

In 1969 'Stingray' became one word. The 1969 model was available with a choice of V8 Chevrolet engines – either the 5.74-liter smallblock or the 7-liter bigblock. The bigblock was increased to 7.4 liters the following year. The 155mph Stingray was to remain practically unmodified until 1983, except that the front was restyled in 1973 with a new, longer nose which incorporated a soft polyurethane cover over the bumper; the rear was restyled the following year too. Sales climbed regularly without feeling the pinch of the oil crisis unduly, and by the end of the 1970s annual sales figures had reached more than 50,000.

The emblem of the Corvette: two crossed flags. One is the checkered race flag, and the other is the Chevrolet company flag. The Corvette, which was a Sting Ray from 1963 to 1967, became the one-word Stingray from 1969 to 1983. The 1968 model was simply known as the Corvette.

detomaso
6676
KX 69
Pantera

DE TOMASO

De Tomaso Pantera

Alejandro De Tomaso was a well-heeled Argentinian racing-car enthusiast who arrived in Italy in the 1950s to satisfy his passion for fast cars. He set up shop near Modena and worked at first in collaboration with OSCA. The De Tomaso firm soon had a reputation for innovative thinking. They were the first to design a power unit to play the role of chassis too, and the first to experiment with a variable rear aerofoil. De Tomaso was quick to cultivate links with the United States, notably with Ford. Carroll Shelby, with his Cobra, had also decided this was the way to go, but De Tomaso had more to offer than an outmoded English sportscar. Any manufacturer who could produce a formidable machine like the GT40 should also be able to offer a modern mid-engined coupe. So De Tomaso offered Ford the Mangusta.

Without a doubt the Mangusta was brilliant, firstly because of its beam chassis based on a load-bearing engine and secondly because it was dressed in what was possibly stylist Giorgio Giugiaro's finest creation. Unfortunately, this superb objet d'art disguised as an automobile was not suited to the production line. Furthermore, Ralph Nader would have had a ball with it. Only 50 Mangustas were ever built.

De Tomaso learned a lesson from this half-failure and soon came up with the more practical Pantera. This time Ford was more interested. The Pantera was built on a robust, selfsupporting monocoque-chassis; the engine was relegated to its old role of providing power (330hp) rather than holding the various components together. The bodystyling was less striking than the previous model's but its compact shape gave a reassuring impression of strength and stability. The 155mph Pantera was built by Ghia in Italy and marketed by Ford in the US at nearly $10,000. Many enthusiasts still mourned the Cobra and sales of the Pantera were sluggish at first. By 1973, Ford was selling more than a hundred each month but the oil crisis put an end to the euphoria. Ford soon lost interest, but De Tomaso still produced the Pantera until the early 1980s.

The overall concept of the De Tomaso Pantera was really the brainchild of Gianpaolo Dallara who came to the firm from Lamborghini, where he had helped create the famous Miura. But the Pantera's engine is no longer a pace-setting V12 but a Ford V8, solid and imperturbable. In the 1973 model shown here, the V8 developed 330hp at 6000rpm and could hurl the 2850-pound Pantera along at more than 155mph. Acceleration was far from sluggish too. The sleek bodystyling seen here was the work of Tom Tjaarda, son of John Tjaarda who had designed the Lincoln Zephyr in the 1930s. The Pantera was built in Italy by the Ghia factory. Alejandro De Tomaso was a major shareholder in Ghia, but he quickly sold his stake to Ford. The Pantera was priced at $9995, a cool thousand more than the Mark IV Continental, the Ford group's most expensive automobile.

FACEL VEGA
2500 3000 3500
100 120 140 160 180 200 220 240 260
80 60 40 20

FACEL VEGA

Facel Vega Excellence

The classical Facel radiator grille is one of the few styling innovations of the 1950s to originate outside Italy or the United States.

The French firm of Facel which became Facel Vega in 1954 was by no means a newcomer to the automobile industry. Facel had already been responsible for the Ford Comet and several Bentleys which needed special bodies. The firm had quickly learned how attractive a classical radiator grille could look on a modern car. The first Facel Vega, unveiled at the Paris Motor Show of 1954, showed this influence. The neo-Bentley radiator was set into the body and given an Italian-style grille. This was flanked by two secondary grilles mounted fairly high. The effect was fresh, original and attractive. The rest of the Facel was reminiscent of the Comet: unencumbered sides, cleanly flowing quarter panels and an elegant roof. But inside (see previous pages) there were some surprises. The dashboard was set out like the instrument panel of a small airliner, with controls like engine throttles. The instruments were round with black dials, which was far from standard practice at the time, and the driver was confronted by an enormous array of switches and instruments. Another innovation, and one that started a fashion which is still with us today, was the driver's central console. This was mounted over the gearbox, which housed several accessories and also served to fill the space between the front seats rather neatly.

On the mechanical side Facel had, for the first time since the war, revived the old established practice among small-scale car manufacturers of importing engines from America. Facel had always gone to the Chrysler group. The time was now ripe for 'going American' again, as the power race had just begun across the Atlantic. The Facel Vegas were sucked into a crazy spiral – they rated 180hp in 1955, 250hp in 1957, 325hp in 1958, 360hp in 1960 and by 1962 they were pushing 390hp! Even the early models were quite capable of reaching 120mph and could reach 60mph from a standing start in under 10 seconds. Unfortunately, the rest of the car, in particular the suspension and brakes, often had difficulty keeping up with the power unit. But this was a failing common among high-performance cars in the 1950s, particularly the Ferraris.

At the 1957 Paris Motor Show, Facel Vega unveiled a 4-door version of its coupe, the Excellence saloon shown here.

Facel Vega unveiled its Excellence saloon at the 1957 Paris Motor Show. The absence of a central pillar between the doors gave the car a certain appeal and also made for easier access to the rear seat, but it certainly did nothing to help stabilize a chassis that was already notorious for its lack of structural rigidity. Yet the undeniable success of the Facel's styling more than made up for this fault. By 1964, 230 cars had been built. The last ones to come off the production line, like the one shown here, had less prominent rear fins, a conventional windshield and a power unit identical to the one under the hood of the HK II described in the following pages.

3249-W-75

Facel Vega HK II

At the beginning of the 1960s, Facel Vega began to go into decline. This small-scale car manufacturer, unlike a majority of newcomers to the automobile industry, had evolved its own particular style right from the beginning. Yet, like its competitors, Facel seemed determined to find its level of incompetence. Even though it already had in production a universally successful luxury car as fast as it was original, the firm threw itself into production of a relatively economical sportscar. This was their first mistake. Since a small car contains almost as many components as a large one it is almost as expensive to build, but the buyer hardly wants to pay as much for it. The producer has to make up for this with mass production, which is feasible as long as one's name is Renault or Citroën. Facel Vega's second mistake was to insist on having a Facel engine as well. Up until then they had always followed the golden rule of choosing an engine which had been tried and tested by a large car manufacturer.

Facel almost succeeded with a car which had an engine of its own. The Facellia, launched in 1959, was well received. It was a very attractive car, offering on a reduced scale the same exciting motoring experience as the earlier models, including the very appealing dashboard. Unfortunately, there was a rather nasty surprise in store for the prospective car buyer. The engine, an attractive 4-cylinder unit with twin overhead cams, had roughly the same life expectancy as a time bomb. The factory was soon devoting much of its energy to replacing engines that were still under warranty – an expensive venture.

Meanwhile, the bigger Facels with their Chrysler engines were relatively trouble-free, although they, too, were showing signs of wear. Nothing goes out of fashion as quickly as something which has just been the rage. By the beginning of the 1960s, no one wanted to be seen behind a panoramic windshield, and exhaust pipes sticking through the bumper bars were definitely part of the 1950s. The name on everyone's lips was Pinin Farina. Buyers wanted harder, more angular styling with sharper ridges and flattened curves.

In its way, the Facel Vega HK II was ten years ahead of its time. With its powerful American V8 squeezed into a relatively compact body, it was a foretaste of the trend toward musclecars that was to sweep America in the 1960s. The 6.3-liter Chrysler V8 came in two versions – the 355hp engine with 4-barrel carburetor and Torqueflite automatic transmission, or the 390hp monster with two 4-barrel carburetors, a compression ratio of 10 and a 4-speed manual Pont-à-Mousson gearbox. The factory's demonstration model with manual gearbox reached 150mph effortlessly and could get from 0 to 60mph in 9 seconds. With the automatic transmission its top speed fell to 120mph, with slightly dulled acceleration, but it had the advantage of power-steering. Fuel consumption in both cases was between 12 and 15 mpg for an overall weight of 3970 pounds. Even though it had been developed about as far as it would go, the big Facel was still a comfortable and safe car in the 1960s. 182 Facel Vegas were built in 1963 and 1964.

So it was that at the 1962 Paris Motor Show Facel launched the HK II. As with all the previous Facel Vegas, it was left to the house designer, Brasseur, to give plastic form to the ideas of managing director Jean Daninos. Between them, the two men came up with a masterpiece. It was as if they had realized that this was their last chance if they were to succeed. The HK II accentuated the innovations which had been introduced on the little Facellia and expanded them on a grander scale: lowered radiator grille, slimmer tail fins, flatter canopy. The passenger was glassed in, with a rear window even bigger than the windshield and slim uprights made of stainless steel (chrome was forbidden on a Facel).

Beneath the new look was the same old Facel Vega: fixed rear axle with leaf-spring suspension, brittle front suspension that had to be renewed about every 10,000 miles, but also disc brakes that had been standard equipment since 1960. The engine was as powerful as ever; it was said that the 390hp Chrysler V8 with its eight carburetor barrels, which could get from 0 to 60mph in 9 seconds, had been developed for the California Highway Patrol. But if the 150mph HK II had reached its logical limits, at least it hadn't been pushed too far. It was a very fast car but had fine roadholding, good brakes and was quite comfortable (at least in the front seats). The firm of Facel Vega could retire two years later with its head held high.

FERRARI

Short Chassis Ferrari

In the Ferraris of the 1950s the engine is definitely king. The bodystyling states this very clearly: the small racing saloons are made up of an outsize engine hood with a cramped little cockpit lost somewhere in the bodywork. It was only in 1959 that Ferrari began to take the needs of the driver into consideration.

At the Le Mans race of that year, Ferrari entered two small saloons of a rather novel design. The comparison between these and the older models, which took part in the same race, is quite instructive. What is most striking is the increased size of the cockpit compared to the rest of the body. The higher canopy allowed more normal windows and a larger windshield. This was the first Ferrari racing saloon to allow the driver any space and comfort. In comparison, driving the older models was like being hunched up inside a broom closet.

In the new saloon a more compact styling was chosen, with reduced overhangs. The V12 at last looked as though it belonged to the same team as the rest of the car instead of seeming to outdistance it. This trend toward integration was reinforced at the Motor Show a few months later when Ferrari unveiled its third racing saloon in less than a year.

This last 1959 model took up the Le Mans design but gave it a wheelbase shortened from 102 to 94 inches. This was the car that was to become famous as the 'Short Chassis Tour de France Ferrari,' and it was so successful that Ferrari kept on making it until 1961.

If there is a Ferrari that marks a turning point in automobile history, this is it. It is not only the first Ferrari, but the first car of any kind to seem just as much at home on the racing circuit as on the suburban street. Certainly, there had already been racing cars, and even Ferraris, that could also be driven on the open road. But they had been first and foremost racing machines. The Short Chassis, with no split-personality problems whatsoever, was just as happy in a grueling race as it was ferrying the kids to school. Once the bugs had been ironed out, the 3-liter V12 turned out 240hp and showed plenty of endurance when handled with a minimum of care. Unfortunately, this balanced personality was shortlived – as the Ferrari saloons evolved they became, more and more, specialized racing machines. This is why the Short Chassis will always have a unique place in Ferrari's history.

The 250GT 3-liter V12 with single overhead camshaft is the mainstay of Ferrari's reputation. The engine under the hood of the Short Chassis saloon, with the new 12-port Testarossa cylinder-head and coil-spring valves, turned out 280hp at 7000rpm.

The Short Chassis small saloon launched in 1959 would have been virtually identical to the Long Chassis Ferrari that came out in the same year, except for two changes that turned it into a totally different car. The stylists had gone back to the drawing board and given it a more spacious cockpit, making several concessions to the driver's comfort such as wind-down windows and a heating and ventilation system. The wheelbase was shortened by eight inches which made the car more compact and much easier to handle. The body, made of light alloy (cr steel and light alloy), was built by Scaglietti, Ferrari's usual body-shop, from a design by Pinin Farina. The car shown here is the 1959 model, recognizable by the absence of ventilation gills on the front wings.

6508

Ferrari California

The name of John von Neumann deserves to go down in history. In 1958 while employed as an agent for Ferrari in Los Angeles, he came up with the brilliant idea of asking Maranello for an open version of the 250GT saloon. Ferrari-lovers should go down on their knees in gratitude – von Neumann gave Pinin Farina and Scaglietti the chance to create a masterpiece.

The prototype was ready by the summer of 1958. Except for minor details (windshield, lateral air vents, bumpers and taillamps), the new car followed the lines of the small saloon, but without the canopy and with the addition of a flat trunk. It was a very attractive car, a cross between a racing shell and a convertible. Its equipment was pared down to essentials like a racer but it had the spacious cockpit of a convertible. The new car was named the California, in honor of the man whose idea it had been. A limited production series was launched at the Scaglietti factory and the first car came off the line in December 1958. In the meantime, the original saloon had lost its streamlined headlamps, so the California had to be modified too. Unfortunately this change reduced the car's appeal. The protruding headlamps lengthened the front wing and the harmonious proportions were lost. One of these cars took fifth place in the Le Mans race, with Ferrari saloons in fourth and sixth places.

In 1959, Ferrari unveiled the Short Chassis saloon described above and it soon had its California version too. For some reason, the short chassis California ended up with completely different lines from the saloon that preceded it. With its headlamps once again streamlined (fickle Italians . . .) and its protruding taillamps, it looks more like a stockier version of the original California. This little gem was christened the Spyder. Its V12 rated 240hp and could get it to 160mph, but one look at it and you can see why it was never a success on the racing circuit. Anyone who laid his hands on one preferred to put it under wraps for posterity rather than risk it in a race.

The V12 under the hood of the Spyder California was the same as the engine in the small, short chassis saloon described in the preceding pages, but it was limited to 240hp at 7000rpm instead of 260hp, and it had a compression ratio of 8.5 instead of 9.5. The Ferrari factory rather optimistically promised a top speed of 160mph with the highest back axle ratio. The V12 250GT had just adopted a new cylinder head which had the sparkplugs located on the outer faces of the 'vee' (on the preceding model they had been on the inner faces). These cylinder heads had been previously tried out on the Testarossa racers. The air vents on the California's engine compartment drew their inspiration from the vents on the Mercedes 300SL.

The short chassis California, vintage 1960, remains one of Ferrari's most perfect expressions of their idea of what an automobile ought to be. In their catalog the makers promised that 'the Spyder California, like the saloon, is just as suitable for touring as for racing.' The California has the same dashboard as the short chassis saloon.

Ferrari Daytona

For the beginnings of the Daytona one should look to the 275GTB which first came off the production line in 1964. That was a watershed year at the Maranello factory. It was the last year the small front-engined GTO saloon was made and the first year of production for its rear-engined successor, the 250LM. However, the new 275GTB unveiled at the Paris Motor Show still had the engine up front – Ferrari had finally made the break between the production-series GT and the competition GT.

In fact, with the GTB it seemed as if Ferrari had allowed time to stand still. The new model certainly introduced innovations, like the rear transaxle and its independent rear suspension, but the engine was once again a derivative of the old V12 designed by Colombo in 1946. Despite the highly fashionable truncated rear section, the Pinin Farina bodystyling, too, could have been found on almost any small saloon in the 1950s. But then, why not?

Ferrari had built a glorious legend and could well afford to retire without loss of face. The outdated Bugattis had still found plenty of buyers in the 1930s. If the factory at Maranello chose to keep producing Tour de France saloons in the 1960s, car lovers were hardly going to complain.

As it turned out, Ferrari was not to rest on its laurels for long. In 1966 the revolutionary Lamborghini Miura burst on the motoring scene like a vehicle from another planet, and of course Ferrari was expected to take up the challenge. When Ferrari launched its twin overhead camshaft 275GTB/2 soon afterward, it seemed like a rather lukewarm reply to the Lamborghini; but any heretical doubts that Ferrari fans might have harbored were finally dispersed when the 365GTB/4 was triumphantly displayed at the Paris Motor Show in October 1968.

As its name suggests, this was basically the 275GTB/4 with the engine capacity increased to 4.38 liters. It was capable of turning out 350hp at 7500rpm in comparison to the 275GTB/4 which turned out 300hp at 8000rpm. The Daytona

The Ferrari Daytona borrowed its powerplant from its predecessor, the 275GTB/4, but increased the engine-capacity to 4.38 liters. On the other hand, its completely restyled body design marked a definite break with the traditional Ferrari look.

Ferrari
Ferrari

4676
KJ 92

engine, a 4390cc four-camshaft V12 fed by six Weber carburetors, proclaimed its presence up front as arrogantly as ever.

If the powerplant was nothing new, the same could hardly be said of the bodystyling. Pinin Farina had made every effort to renew his art, even to the point of risking Ferrari's traditional image; the construction of the bodywork was left to Scaglietti. In styling, the 365 broke completely with the 275. The earlier car's voluptuous curves gave way to a brutally sober machine stripped for combat. The aim was to rival the aggressive modernism of the Miura.

While still at the prototype stage, the car had been christened the Daytona in honor of the Ferrari successes on the American circuit, and the name stuck. The Daytona's performance gave nothing away to its Lamborghini rival, with a top speed of 170mph. But if the Miura was above all a flamboyant sportscar, the Daytona was basically a powerful freeway cruiser. Admittedly the Ferrari was hard to maneuver in tight corners, but it made up for that with its superb stability at high speeds. The driving position was hardly inviting; it was often compared to the cabin of a big truck because its enormous steering wheel was too high and too flat. The windshield was too far from the driver and prevented him from even seeing the engine hood.

Yet these minor inconveniences must have seemed a small price for the prospective owner to pay – it could justifiably claim to be the fastest car on the road – for the advantages of the superbly purring 4.4-liter V12 up front.

Like many rivals, the Ferrari Daytona and the Lamborghini Miura in fact made the ideal couple. They had very similar performance characteristics but the ways they went about achieving these were very different. On their dashboards, the dials bore the same figures, but they were presented in very different ways – an almost banal classicism in the Daytona and a rather 'way out' originality in the case of the Miura, with its switches on the cabin roof. Today, the Ferrari has developed a strong cult following, which is more than can be said for its Lamborghini rival. Justice may not have been done. But then the Daytona now belongs well and truly to the past and lends itself more to nostalgia. The Miura still seems as modern today as the day the first car rolled out of the factory.

100 140
TEMP OLIO
ACQUA
40
50
60
70
80
VEGLIA BORLETTI
GIRI x 100
0 40
AMPERE
5 10
Kg/cm²
PRESS. OLIO
12
3
6

Ferrari 512BB

When Ferrari-lovers talked about BB in the 1970s, there was a good chance that they were talking not about Brigitte Bardot but about the Berlinetta Boxer. Boxer is a German term used to describe engines with 'flat,' horizontally-opposed cylinders, in which the pistons seem to box against each other. In other words, according to Maranello terminology, the code 512BB meant 2-seater, 5-liter, flat 12-cylinder coupe.

This was the model that came out in 1973 to take up where the Daytona had left off. At first it was called the 365GT/BB under the old code reserved for V12 cars (as well as straight 4- and 6-cylinder models) indicating engine capacity: 365cc (81mm x 71mm) or 4.4 liters. Power was rated at 380hp at 7200rpm (the Daytona turned out 350hp at 7500rpm).

What these statistics didn't reveal was that the cylinders were horizontally opposed and that the engine was at the rear of the BB. This may have been a surprise, but it wasn't about to shock the old-school Ferrari connoisseurs. They were well aware of how much progress had been made since 1964 by the flat twelves in competition models, and the arrival of the Dino in 1968, with its mid-engine, had made them realize that a new age had dawned at Ferrari. In fact, many of them had taken the Daytona as an unexpected reprieve, a final gesture to the old guard.

When it succeeded the Daytona, the BB made no claim to being a super-Dino. Like its predecessor, it preferred stability at high speed to agility in tight corners. Drivers who were used to the Daytona were in fact more at home behind the wheel of the BB than those who came to it from the Dino. For the Daytona men there was no new dimension: the BB had a top speed of 170mph. Progress remained modest and the magic figure of 185mph was still out of reach.

The body for the BB was built by Scaglietti to a design by Pinin Farina. The BB had clean, balanced lines, with a tense feel about them that was a welcome change after the sloppy curves of the Dino. Pinin Farina hadn't taken any risks but hadn't displayed much originality either. It is true that the Lamborghini Countach was already on the scene and its futuristic styling would have been hard to beat. Yet in 1971 Pinin Farina had built an astonishing, even more radical prototype (with a Ferrari engine to boot) called the Modulo, that could easily bear comparison with the Countach. Strangely enough, the BB bore very little

Coming two years after the Countach, the Berlinetta Boxer did not attempt to compete with the Lamborghini in the field of styling. But the BB's classical aura betrays an undercurrent of tension which suggests that it was hard-won. Two years before the Boxer, Pinin Farina had shown off his extraordinary Ferrari Modulo, an experimental prototype that gave nothing away to the Countach in futurism.

Ferrari

Ferrari
765 WWA 92

The Berlinetta Boxer owed its name to its horizontally-opposed, 12-cylinder engine, a descendant of the powerplants designed by Mauro Forghieri for competition cars. The first version, released in 1973, had a capacity of 4.4 liters. It developed 380hp at 7200rpm with four triple-throated Weber carburetors. The BB had a top speed of only 170mph, which disappointed those who had expected it to reach the magic figure of 185mph. The Ferrari faithful hoped for better things to come, but in fact the opposite was to occur. In 1976, the engine capacity was increased to 5 liters but at the same time the power was reduced to 360hp in order to satisfy the new American safety standards. Meanwhile, the Boxer had put on 650 pounds in weight. In 1982, the BB was equipped with a Bosch fuel-injection system, reducing its power even more to 340hp. The 1982 model is shown here.

trace of the adventurous prototype. Its classical lines were undoubtedly the result of a precise marketing strategy. Ferrari had no real desire to engage in open competition with Lamborghini. Furthermore, Ferrari had gotten Bertone to design the body for their 308GT4 which replaced the Dino and which was released at the same time as the BB. The result had been quite disappointing – with its rough forms, the 308GT just served to underline the elegance that had been achieved by Pinin Farina.

When it was launched, the BB provoked little enthusiasm. It never quite managed to make its mark and laid the ghost of the Daytona. This was partly due to the oil crisis and a tightening of technical standards in the United States. The engine capacity was soon increased to 5 liters (82mm x 78mm), which made it the 512BB, perhaps meaning that Ferrari did not quite regard it as worthy to be included in the royal line of V12s. Its power dropped to 360hp at 6500rpm – a drop of 20hp, although the speed was increased to an estimated 188mph – and at the same time the addition of accessories (air-conditioning and electric windows) raised its weight from 2650 pounds to 3300 pounds, which didn't do much for performance. The car's decadence was reinforced in 1982 when Bosch fuel injection replaced the four Weber triple-throat carburetors and reduced its power still further to 340hp at 6000rpm. In 1984, the BB finally made way for the Testarossa which sought to bring the Ferrari legend back to life.

Ferrari
Ferrari s.p.a.
COSTRUITO DA:
ufi UNIVERSAL FILTER ITALIANA s.p.a.
Ferrari
Agip
MOTORE Agip SINT 2000 SAE 10W/50
DIFFERENZIALE CAMBIO Agip F.1 Rotra MP SAE 80W/90

FORD

Ford GT40

During the summer of 1962 Henry Ford II announced his 'Total Performance' project – Dearborn had decided to attack the racing scene on all fronts, from Indianapolis to Le Mans. This was also a prelude to the Mustang's attempt to capture the postwar 'baby-boom' market.

In mid-1962, the famous Mustang was still a rear-engined prototype, the brainchild of 'imported' British engineer Roy Lunn. In the end, Lee Iacocca was to reject this first Mustang as too expensive for mass production, but the car was to serve as the basis for a Grand Tourisme class competition model built around a 4.2-liter rocker-arm Ford V8 and intended for Indianapolis. During the rest of 1962 the design was modified bit by bit. In the end, the open racer was roofed over: Ford had decided to produce the most comfortable racing car ever built. After all, at Le Mans a driver has to sit behind the wheel for three or four hours. In addition, a coupe lends itself to more efficient aerodynamics than an open racer. Ford had in mind a top speed of well over 185mph, which is the speed at which a jet fighter leaves the ground. To stop the Ford from taking off too, its body design was subjected to rigorous wind-tunnel testing. The radiator was moved to the rear, with an air extractor on the hood taking advantage of a slight depression. The various air intakes for supplying cabin ventilation, brake cooling, and carburetor feed were also designed with great attention paid to aerodynamic flow. The racer's body was to be built of fiberglass, which was well adapted to a limited production run. The body would sit on a selfsupporting platform chassis with enormous welded caissons acting as fuel tanks. As the driver would have to clamber over these to reach his cockpit, Ford even considered hinging the doors in the middle of the roof, like the 'wings' on the Mercedes 300SL. As this would end up trapping the car's occupants inside if the car overturned, the access doors were finally hinged in the standard way, but with a large slice of roof included as part of the opening. When the time came to choose a name for the car, Ford settled on the GT40 as the car stood 40 inches high.

The GT40 had set the standard features not only for the Ford competition coupes but for all the other mid-engined GT coupes that were to be so prolific in the 1960s. The GT40 was the idea of the moment – at the London Racing Car Show in January 1963 the Lola GT went on display. It was a small, mid-engined saloon built around the same Indianapolis Ford V8. Dearborn immediately engaged its creator, Eric Broadley, hiring at the same time the manager of Aston Martin, John Wyer. These two men were soon joined by Roy Lunn. Together they set up the Ford works at Slough, on the outskirts of London, and from 1965 they began turning out the one hundred powerful (385hp at 6000rpm) GT40s necessary to meet class requirements.

This is where it all began: the small mid-engined sportscar that was to dominate the 1960s was born with the Ford GT40. The car went on display for the first time in New York in March 1964. Its success was due to the combined efforts of Ford and Eric Broadley, the man behind the Lolas. During development of the GT40, particular attention was paid to the car's aerodynamics; its efficient shape was only achieved after many initial difficulties. The GT40 still remains one of the most comfortable racing cars ever built, yet it was above all a fighting machine, ill-suited to touring. The Ford V8 engine in the 1966 model shown here developed 385hp at 6000rpm for an all-up weight of 2400 pounds. Positioning of the gear lever makes it quite clear that this is a competition car and not one of the road cars built in 1967.

C&R
Clothiers

JAGUAR

Jaguar XK 120

In early 1948, the men at Jaguar were hard at work on a luxury car they hoped would become the firm's standard bearer in the coming years. Its creator, William Heynes, had insisted on an engine with dual overhead camshafts, which was most unusual in a luxury car. It was a very British 3.4-liter (83mm x 106mm) 6-cylinder engine which was required to be, above all, quiet and respectable. The car was named the Mark VII Jaguar, partly because one of its tasks was to outdo the Mark VI Bentley. However, Jaguar never intended the Mark VII to be a sportscar. The British economy badly needed dollars and everyone knew that no one was buying sportscars in the United States.

The new chassis was ready. All that was missing was the body that had been subcontracted to a specialist. The Mark VII was going to steal the thunder at the first Earls Court Motor Show held in London since the war. Then a stroke of bad luck seemed to dash all Jaguar's hopes. The bodywork contractor proved unable to meet its deadline. William Lyons, the iron-willed Jaguar director, rose to the challenge. The new Jaguar had to go to Earls Court whatever the cost. If the chassis alone was ready, then the chassis would go on show. However, to be sure of making the news-

The Jaguar XK 120 is a masterful combination of features copied from other cars – there is a little of the Bugatti 57S, a touch of Figoni's Talbot, and a dash of the Mille Miglia BMW. Jaguar's chief, William Lyons, took personal charge of designing the bodywork of his cars. The XK 120 was based on a shortened version of the Mark VII saloon's frame and its handling characteristics benefit from this. The early cars (some 243 in all) had a light alloy body built on a wooden framework. Their engine developed 160hp and their top speed of 120mph did justice to their codename XK 120. The later cars, with steel bodies and power reduced to 150hp were not as fast. This led Jaguar to market special equipment (pistons, camshafts, exhausts etc) derived from the Le Mans XK 120C ('C' for Competition) that could bring the engine's power up to 180hp. In all, 12,298 of the XK 120 series were produced before the model gave way to the XK 140 in September 1954.

papers, Lyons would have the chassis dressed up, quickly and cheaply, as a spectacular open 2-seater sportscar. It would only have to last as long as the show, as there wouldn't be any buyers for a car like that in depressed, postwar Britain.

Lyons himself took charge of the bodystyling. Having no particular bent for the plastic arts, he had a series of full-scale mockups built and then had them modified until he was satisfied with the result. As he had good taste and was responsible to no one, the result was splendid. All that remained was to find a name for Lyons' creation. The twin overhead cam, 6-cylinder 160hp engine belonged to a family of experimental engines designated XA, XB, XC, etc, and the one now under the hood was the XK. The car was expected to have a top speed of 120mph, and when prewar Jaguars still had the SS label, SS 100 designated a model with a top speed of 100mph. The new car would just have to be called the XK 120. It was to have quite unprecedented success on both sides of the Atlantic.

9751AL92

Jaguar XKE

With its 1961 XKE, Jaguar managed to steal the show for a second time just as it had done with the XK 120. Once again the firm could offer unrivaled performance for about half the price of an Aston Martin or a Ferrari. Indeed, the XKE was an even more impressive achievement than its predecessor. The XK 120 had been a brilliant improvisation but still bore the traces of the haste with which it had been put together. The XKE was thoroughly planned and completely justified its boast of being the 'avant-garde sportscar.'

As its name indicated, the XKE was a descendant of the competition D-Type. There had never been an A- or B-Type, only the XK 120C ('C' for competition) which was so widely known as the C-Type that the next car automatically became the D-Type. The C-Type Jaguar or XK 120C was a revamped version of the XK 120 with new rear suspension, tubular chassis, rack-and-pinion steering and a more functional body. The 120C, possibly Jaguar's masterpiece, won Le Mans in 1951 and 1953. Jaguar then put the 120C on sale at a price no higher than the Aston Martin DB2.

The D-Type which followed, went a little farther along the same path. The tubular chassis was retained in front, but aft of the engine bulkhead a light alloy, selfsupporting shell took over, saving about 220 pounds. The bodystyling was reminiscent of the Alfa Romeo Disco Volante but had an air scoop included up front. The D-Type, produced to win at Le Mans, was an immediate success, winning the race for three years in succession, in 1955, 1956 and 1957.

The D-Type Jaguar had a fixed rear axle, but this last archaic feature was done away with on a very special Jaguar entered by Briggs Cunningham for the 1960 24-Hour Race. Other differences were a slimmer body and independent rear suspension with single cross-arms and two pairs of coil springs. The disc brakes were sighted right next to the differential. This model brought together the main ingredients of the XKE. All that remained was the official launch of the new model, which took place the following year at the Geneva Motor Show.

When the Jaguar XKE's hood (which means a good third of its body) is raised, the rather peculiar chassis construction is revealed: a selfsupporting shell married to a tubular frame at the level of the engine bulkhead. This is a marriage of reason, combining lightness with rigidity. The value of the technique was proved by repeated successes in the Le Mans 24-Hour Race. The XKE might be described as the XK 120 improved and updated in the light of racing experience. The original model provided the engine and the front suspension. The rack-and-pinion steering comes from the C-Type, the monocoque chassis from the D-Type, the rear suspension from Cunningham's 1960 Le Mans racer. The original version which appeared at the Geneva Motor Show in early 1961 had a 3.8-liter engine like the car shown here. That was the most brilliant model of the series but it was badly let down by a mediocre gearbox. Thereafter, the XKE was offered in a choice of two versions: the 2-seater convertible shown here and an Aston-Martin-style hatchback coupe.

Instead of the Le Mans racer's light alloy engine with Lucas fuel injection, the 1961 XKE Jaguar received the engine that already powered the XK 150S, with three SU carburetors and an engine capacity of 3.8 liters (87mm x 106mm) which could develop 265hp at 5500rpm. The makers promised a top speed of 150mph and the demonstration car that was lent to motoring writers reached precisely that speed. Even though the production models sold to the public rarely exceeded 140mph, the XKE was one of the best deals that Jaguar ever offered its customers. The motoring magazines of the day were full of praise for its comfort and its roadholding, not to mention its light yet precise steering. The brakes were sometimes judged inadequate for the car's power but the gearbox was roundly condemned. Already out of date by the 1950s, it was anachronistic to keep using it in the 1960s. Only in 1964 did the XKE get the gearbox it deserved, while its engine capacity was increased to 4.2 liters. Since the power output was unchanged while the car had quietly gained weight, its performance suffered. Decadence really set in when the XKE lost its streamlined headlamps and then subsequently its wire wheels.

In the end the XKE, as an 'avant-première,' was given the V12 engine destined for the future XJ 12. However, this monster, like the masterpiece it powered, was soon done to death by American antipollution standards.

LAMBORGHINI

Lamborghini Miura

In automobile history, there are the GTs that come before the Miura and those that come afterward. Before the Miura, the engine was in front of the driver. As engines became more powerful, hoods became longer, rather like the nose of a piston-engined fighter plane. Driving the Miura was like sitting on top of a rocket. The engine was behind the cockpit where it could take up all the space it needed without restricting the driver, who sat low down behind a nose panel which had been reduced to aerodynamic essentials.

Everyone was waiting for the Miura in the 1960s. Racing cars had already adopted the mid-engine and the industry needed an exciting tourer along the lines of the Ford GT40. Yet Ferrari, Maserati and even Aston Martin and Jaguar seemed in no hurry to take the plunge. They had all built their reputations with a certain type of car and did not want to risk damaging their prestige by moving in a radically different direction. They were probably right. Car fanatics tend to be very conservative – witness the faithful who will probably never forgive Ferrari for the Daytona.

Certainly no one in the car world was expecting the Miura to come from Lamborghini. The Sant'-Agata car manufacturer had often enough been scornful about the racing scene. For Lamborghini, all modern GTs suffered from being badly assimilated racing machines. So they settled for producing cars which combined both the performance of a Ferrari and the refinement of a Rolls-Royce. Given its stated aims, Lamborghini was unlikely to be the one to adopt the latest innovations spawned by the rigors of competition.

When the 350 GT was released in 1964 it confirmed this impression. The car was certainly fast, but it lacked character to the point of being downright ugly. But not everybody at Lamborghini shared their chief's vision of what a car ought to

According to the 1900 Baedeker Guide, Signor Miura was the breeder of some of the most renowned fighting bulls in Europe. But he is famous today above all because he gave his name to the Lamborghini. The Lamborghini Miura was dreamed up by a group of men, all under 30 years of age, who were unhampered by tradition and prejudice while working for a car maker as yet unconstrained by the need to protect an established image. The cat-like bodystyling (see photo previous pages) conceived by Marcello Gandini was the perfect complement to the audacious chassis built by Dallara. The Lamborghini V12 had brought about a double revolution by being placed in the rear of the car and pivoting 90 degrees onto its side. The original 1966 version of the Miura had an engine capable of 350hp. This was raised to 370hp for the Miura S and finally pushed to 385hp for the Miura SV. The car's top speed varied from 170 to 180mph. Between 1966 and 1972, 765 Miuras were built. 149 of these were the SV version shown here.

be. The engineers Dallara, Stanzani and Wallace, whose average age was only 25, dreamed of building a true racing car. They got together in a corner of the factory and built a racing chassis with a V12 engine mounted sideways. Confronted with this 'fait accompli,' Ferruccio Lamborghini gave the project his blessing, but he would still not hear of a racing car. It would be a tourer.

When the chassis was displayed at the 1965 Turin Motor Show, it provoked the mixture of admiration and skepticism that usually greeted Lamborghini's creations. But as soon as Bertone saw it he wanted it for himself. His timing was just right. The firm of Touring, Lamborghini's usual body-builders, had just gone bankrupt. Fate had favored the Miura a second time and Bertone was much better qualified than Touring to create the flamboyant body that would match the new car's exciting specifications.

The next episode in the Miura story took place at the 1966 Geneva Motor Show. The 350hp, 170mph Miura on display under the Lamborghini banner inspired such enthusiasm that its creator had to abandon all his dreams of a super-civilized GT and rush into production what was to become one of the most aggressive (the 385hp SV model shown here was capable of 180mph) and most radical small saloons in automobile history.

Lamborghini Countach

The Countach succeeded the Miura in much the same way as the A-bomb led inexorably to the H-bomb. Everyone thought that the Miura was the ultimate in cars, and then Lamborghini came up with an even more devastating machine. This was at the Geneva Motor Show in 1971. Yet today the Countach seems even younger than ever.

If one had to find fault with the Countach, then it would be for bringing the Miura's career to a rather premature end. At that stage, Ferrari was still touting the Daytona, and the Boxer was still two years away. The SV version of the Miura had appeared on the scene and easily outclassed all the hopeful competitors. Only Lamborghini could do better than Lamborghini.

At first sight, the Countach seemed to have taken a retrograde step in its design when compared to its predecessor. This is because once again the engine sat fore-and-aft. But the retrograde move was an illusion: in reality the power-unit had turned a further 90 degrees to complete an about-face. The usual engine position had been reversed in the Countach – the gearbox was now placed forward of the engine, with a transmission shaft traversing the V12 crankcase to reach the differential mounted at the other end. This arrangement enabled the designers to sidestep two pitfalls – the excessive length of those power units where the gearbox is located between the engine and the differential; and the overhang, detrimental to roadholding, which results when the gearbox is placed aft of the differential. The innovation also brought an additional advantage – the driver quite literally had the gearbox at his fingertips, thus affording him the most direct and precise control possible.

Today, just as in 1971, the Countach is remarkable first and foremost for its body which is truly stunning. It would be nice to be able to find another word for this piece of machinery which is a cross between a minisub, a war machine and a spaceship. It is a sort of mechanical totem, half-barbaric and half-sophisticated, which appeals to our innermost fantasies.

The Countach was shaped by Marcello Gandini, Bertone's chief designer, and bears the traces of two of his previous creations, the Carabo and the Stratos. The doors pivot upward like those on the Carabo. The hood comes from the Stratos. It is a direct extension of the windshield and does nothing more than conceal the spare wheel and the driver's feet. But if the front of the car seems to have become atrophied, the rear has grown to outsize proportions. The rear window resembles the driver's slit of an army tank and the taillamps look as if they are about to emit a death-ray. The radiators are arranged at the rear too. Their air intakes, almost invisible on the prototype, have protruding scoops and extractors.

The first 'production series' Countach went on show in Geneva in March 1974. There were several important differences between this and the original model. The engine capacity had been reduced from 5 to 4 liters. The welded-caisson chassis, rather like the Miura's, had been replaced by a tubular steel frame which was heavier but less expensive to produce. These changes were wrought because Lamborghini had been in financial difficulty. The little Uracco, similar in concept to the French Facellia, had not enjoyed the success which had been hoped for. Ferruccio Lamborghini had been forced to sell out to foreigners, and from that moment the firm changed hands almost as often as Italy changed governments. But by then the Countach had achieved a momentum of its own and the idea of stopping production was inconceivable. In 1978, the S version was released, with widened wheel arches for the new Pirelli P7 tires. This car could reach a speed of 60mph in under 5 seconds and 120mph in under 18 seconds. In 1982, the engine capacity was brought up to the original 5 liters. The power developed was unchanged at 375hp but the engine had become more docile. The Lamborghini Countach shown here was manufactured in 1984.

With its tapering nose, outsized rear section and air intakes like those on a supersonic fighter, the Lamborghini Countach seen from side-on looks rather like an F15 after a wheels-up landing.

Seen in the rearview mirror, the Countach S is a sight liable to make a driver's heart skip a beat. Yet according to the experts, it is one of the safest and most docile Lamborghinis ever to roll out of the Sant'Agata factory. Its trajectory is as smooth as a passenger jet's and yet it is as forgiving as a well-worn Citroën 2CV. This stunning machine won't quite reach 180mph, particularly with the optional rear spoiler which guarantees that the Countach won't leave the ground – but at the sacrifice of 10mph. The car shown here is a 1984 model. Since 1975, some 800 Countachs have been built, half of these being the 5-liter S model which has been standard since 1982.

F-GAOA
5433 LE 92

LANCIA

Lancia Stratos

The Lancia Stratos took its name from an extraordinary prototype presented by Bertone at the 1970 Turin Motor Show. It looked as though it came from outer space.

To get some idea of what it was like you have to take the version shown here and mentally eliminate the protruding cockpit, leaving only the main fuselage. The occupants had to lie almost prone under a windshield that doubled up as hatch. This was automobile bodywork reduced to the bare essentials, and one of the most radical creations of Marcello Gandini, who had already put his signature on the Carabo and was soon to give the world the Countach.

Contrary to all appearances, the Lancia Stratos really was an automobile, and it really went. The chassis was an original design by Bertone. The power unit was borrowed from the Lancia Fulvia and was mounted centrally at the rear.

As luck would have it, Lancia was at that stage looking for a car to take over from the Lancia rally coupe which had had its day. In spite of its magnificent exravagance, the Bertone prototype seemed to be exactly what they had been looking for. As soon as they learned that the roadtesting of the Stratos had been satisfactory they contacted the bodybuilder. They already knew what they were looking for – a mid-engined fairly compact coupe with a short wheelbase, which would be able to outdistance the rival Porsches and Renault Alpines. In its competition version the Fulvia engine had been brought up to 160hp and had gone about as far as it could go. But for the preceding year, Lancia had been part of the Fiat empire, which also controlled Ferrari . . . The sideways V6 in the Dino 246GT fulfilled all the necessary requirements. So Ferrari would provide the engine, Lancia would design the car and Bertone would dress it up.

The original design of the Stratos was completely reworked to fit the new conception of the car. The cockpit popped up out of the body complete with windshield and lateral windows worthy of the name. Gandini being Gandini, the windshield had a novel, wraparound shape with

Seen from side-on, the Lancia Stratos bears Bertone's distinctive signature. The similarity in styling between the Stratos and the Lamborghini Countach or the Ferrari 308GT4 is obvious. Yet the wraparound windshield and the short wheelbase remain the trademarks of this compact little machine. The Stratos rarely managed to retain its clean, flowing lines. Designed above all to be agile and maneuverable, stability at high speed was not the car's strong point, and additional aerodynamic accessories very soon proved indispensable. 500 of the ordinary customer version shown here were produced in 1974. They were powered by the same 2.4-liter V6 engine as the Dino Ferrari 246GT. With the engine developing 190hp at 7000rpm for a low weight of 1900 pounds, the top speed of the Stratos was over 140mph. Seen from a three-quarter rear view (center right), the Stratos looks more than ever like a space capsule which has just released its booster rockets. The first Stratos (top right), styled by Bertone, was presented at the Turin Motor Show in 1970.

contours flowing up into the side windows. The front of the car ended up looking much like the cockpit of a Learjet, the airplane renowned for its sensational performance and its low-slung cabin. The prototype of the Lancia Stratos was one of the star attractions at the 1971 Turin Motor Show. It was interpreted as a sign of the firm's rebirth after the Fiat takeover. The racing version was finished a year later.

Further development went on throughout 1973, supervised by engineer Gianpaolo Dallara who was already famous for the work he had done with both Lamborghini and De Tomaso. By the end of the year, the Stratos was deemed ready, and Lancia started production of the five hundred cars necessary to obtain a GT rating from the FIA. Production began in November 1973 with the first target of homologation, 400 cars, reached in 1974.

The chassis and body were supplied by Bertone to be assembled at the Lancia factory. Despite their 140mph top speed, the 2.4-liter V6 machines proved difficult to sell and they remained in the Lancia catalog for four years. During this period the Stratos won the World Rally Championship three times, in 1974, 1975 and 1976. The competition models developed 240hp (the customer model 190hp at 7000rpm), which could be raised to 340hp by using a turbo. In the end, despite its mixed blood, the Stratos brought honor to the name of Lancia.

MASERATI

Maserati Ghibli

Toward the end of the 1960s, the Italian automobile industry went through its equivalent of the Renaissance. Swapping of ideas among mechanical wizards, designers and bodystylists produced a magnificent display of automotive masterpieces. During this period of creative flair it was inevitable that some of the less spectacular achievements would fall by the wayside and never receive the recognition they deserved.

That was certainly the fate of the Maserati Ghibli. The Ghibli belongs, in time, between the Lamborghini Miura and the Ferrari Daytona, and seems like a classical symphony coming between two rock concerts. It had balance, refinement, and above all simplicity. The flat surfaces are relieved by the merest whisper of a curve; the ridges are clean but with the sharpness removed. The 4.7-liter V8's hood gives an impression of flatness but is in reality a subtle combination of convex and concave curves. One can sense the powerful 330hp muscles rippling under the Ghibli's skin, but it is the power of a well-trained animal that will do exactly as it is told.

If the role of a car body is to express in three dimensions the personality of the vehicle, then the Ghibli's styling is a masterpiece, just as much as that of the Daytona or the Miura. The Ghibli is a perfectly conceived car and admirably docile; it remains just as controlled whether it's traveling on a highway at 160mph or cruising at a more leisurely speed on a Sunday afternoon drive.

The dictionary defines ghibli as a hot, dry wind which blows across the Libyan desert. Yet the Maserati Ghibli is no mere blast of hot air. It is a superbly civilized car, possessing an exotic touch for good measure. That is exactly what engineer Giulio Alfieri, Maserati's project chief, had in mind when he developed the car. With its solid rear axle, its big twin overhead camshaft V8, its heavy weight (3300 pounds) and the timeless classicism of its styling, the Ghibli is like an Italian-style Aston Martin. Its personality was admirably expressed in plastic form by bodystylist Giorgio Giugiaro, who had come to Ghia at the beginning of 1966 after working for Bertone for six years. Giugiaro was just 26 and in top form. The Maserati Ghibli is perhaps his most perfect creation.

The impressive Maserati V8 has a capacity of 4.7 liters (94mm x 85mm) and develops 330hp at 5000rpm, turning out a top speed approaching 160mph. The only carburetors thought suitable for such a pedigreed Italian car were dual-throat Webers. With very energetic torque at low revolutions, the engine's suppleness was an important factor in ensuring comfort for the Ghibli's driver. As an added luxury he had a well laid-out cockpit and air-conditioning too.

428 DHH 75

Maserati Bora

The shock wave from the Miura rocked Maserati just as it rocked all other high-performance car manufacturers. The consequences were inescapable, with the expected new models soon appearing. In this competitive climate Maserati unveiled *its* mid-engined coupe, the Bora, at the 1971 Geneva Motor Show.

The bora is a north wind. If names are anything to go by, then the car world probably thought that after the burning wind of the Ghibli the Bora would be more temperate. This was not the case, for the good reason that it was still the same wind, and still the same car, even if the priorities had been rearranged.

The engine is still a deep-throated V8 with a rather prosaic tone at low revs. Fitted with four twin-choke Weber carburetors, it still develops 330hp at 6000rpm. In the Bora the 4.7-liter V8 is mounted lengthways, in line with the ZF gearbox. This means that the engine compartment manages to take up a good bit of the car despite the imposing, glassed-in portico that Giorgio Giugiaro has stuck on top.

The priorities have changed, but the car is nevertheless basically the same as the Ghibli – it is a seriously thought-out car, the very antithesis of a brilliant improvisation. Firstly, the Bora has one quality that is quite unexpected in a mid-engined GT, and that is silence. The whole of the rear power unit, and that includes the suspension, is fixed to a tubular cradle (something like the tubular framework supporting the engine of an XKE) which is joined to the forward monocoque chassis using damper blocks. This effectively filters out any extraneous noises produced by the engine. The Maserati V8 offers its driver a less impressive performance than the V12 in the Miura. Roadholding in the Maserati is good, but the car has a noticeably firm ride.

One sign of the times was that the Bora was equipped with a Citroën 'hydraulic unit' – a hydraulic pump with an accumulator, linked to the braking circuit by that ultra-sensitive pedal that Citroën owners know so well. At that stage the French car manufacturer was being a little heavy-handed with Maserati who were providing the engines for the SM. The hydraulic pressure in the Bora also helped the driver to adjust the pedal position to his own tastes.

Like the body of the Ghibli, the styling of the Bora is signed Giugiaro. But the years had gone by, and Giorgio had left Ghia just as he had left Bertone; he had gone out on his own by founding Ital Design. The lines of the Bora still show an undeniable kinship with those of the Ghibli, and this was possibly Maserati's intention. The radiator grille in particular is an overt allusion to the past. In general, the Bora's lines are more supple, less sharply cut than most of the work which Giugiaro had been doing in the early 1970s.

In the end, his creation for the Bora proved its remarkable vitality by surviving until recently in the body of the Merak.

The Maserati Bora was a truly original attempt at a mid-engined GT, with the accent on comfort, silence and ease of driving rather than on more sporty characteristics. The Bora borrowed from the front-engined Ghibli not only its main mechanical components but also the traits of personality that made it a sort of 2-seater executive jet designed for the freeway. So its natural rival ended up being the Citroën SM rather than the Lamborghini Miura. In other words, the Bora's basic configuration didn't have a lot going for it even though the car ended up being a success in itself.

MERCEDES

Mercedes 300SL

It is 1951. Mercedes is preparing a racing comeback for the following season and is faced with a problem. The cars are all streamlined coupes and it's impossible to get in and out of them. The chassis is a tubular carcass that takes up the entire height of the body on both sides of the cockpit. You can't cut into that to provide a door or the whole body might collapse. Then engineer Rudolf Uhlenhaut has an idea: the car is very low-slung, so why not get in through the top? All you will need are two doors hinging in the middle of the roof and incorporating the side windows. To get into the car you lift the whole thing and slip down behind the wheel. To make things easier for the driver, the steering wheel tilts forward. At the last moment, Mercedes adds another refinement. The sides of the chassis are lowered somewhat, so the wing doors will also be able to include a chunk of the side panel.

The new 300SL (3 liters, Sport Leicht) impressed the Italian public enormously at the start of the Mille Miglia, particularly as the cars were so badly ventilated (their side windows were fixed) that the drivers sometimes allowed themselves the luxury of driving for a while with the wings up. So by overcoming these technical problems Mercedes made a truly sensational styling discovery.

The 300SL had an excellent season in 1952 – second in the Mille Miglia, first both at Le Mans and in the Pan-American. The success in the Mexican race attracted attention in the United States. Daimler-Benz's agent, the famous Max Hoffman, succeeded in persuading his head office to market a touring version. This was unveiled in early 1954 and held several surprises. Firstly, the body had been restyled with more generous curves around the hips and shoulders. But the car's performance proved even better than that of the racer. The 300SL had traded in its Solex carburetors for the new direct fuel-injection system developed by Bosch, which was being tried out on the Mercedes even before the Formula 1 cars. This was the first time ever that a car manufacturer had been able to produce a model capable of breaking what had been for 20 years the tourer's 'sound barrier' – 120mph.

Despite appearances, the 300SL had the same wheelbase as the little 190SL – 94 inches, but the rather restricted cockpit makes the rest of the body look bigger. The surprising thickness of the quarter panels, like the unusual door configuration, was made necessary by the tubular lattice construction of the chassis. The mechanical side was borrowed in the main from the 300S saloon, starting with the 3-liter (85mm x 88mm) straight-six overhead camshaft engine. With a little help from the Bosch fuel-injection system, the 300SL could develop 240hp or even 265hp depending on the compression ratio (8.5 or 9). Not bad when you consider that the tourer could only manage 150hp and the carburetor-fed competition model only 185hp! In theory, the top speed was nearly 160mph but in reality it was closer to 135mph. Even this was an unheard-of figure in 1954 for a production car, which made it a little easier to forgive the 300SL the rather fragile roadholding caused by its swing-axle suspension. Because of the fixed windows, ventilation for the car had to be very carefully engineered, with air extractors installed just above the rear window (see photo opposite). And finally, the bad news: the 300SL cost more than $7500 at a time when the 2CV cost $500.

CH
SG
2019

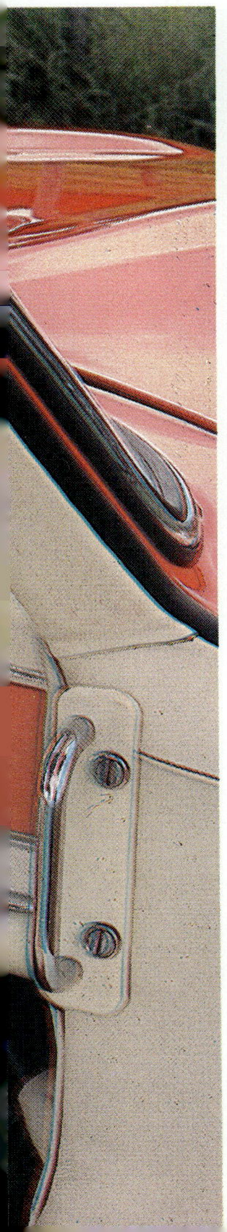

PONTIAC

Pontiac GTO

In the early 1960s some car designers in the United States began to notice that European car manufacturers had a knack for designing high-performance cars around the good old V8s which were being used in America for driving Mom and the kids to the supermarket. So American designers decided to try the same technique which would provide an outlet for the postwar 'baby-boom' teenagers who were just learning to drive. At that stage, Pontiac was working on a new Tempest which, like its predecessor, was a 'compact.' This meant that it was a large enough saloon car but was powered by an inoffensive 3.5-liter straight-six-cylinder engine. The full-size Pontiacs were the models in the Catalina-Bonneville range, powered by the 6.4-liter V8. Why not put the V8 in the compact, thought the chief engineer at Pontiac, an ambitious young man called John Z DeLorean? For good measure, the engine's power was boosted to 348hp with three dual-throat carburetors and a compression ratio of 10.8 to 1 (or 325hp with a single 4-throat carburetor). At the same time, the body was stripped down to its bare essentials. Pontiac ended up with a Tempest whose rear wheels could almost catch up with its front.

This model was christened the Pontiac GTO – a sly wink at the Ferrari series of Gran Turismo Omologato saloons, so-called because their right to the GT rating had been long contested. Strictly speaking, the new Pontiac was not a new model at all but an optional V8 version of the old one comprising, in addition to the engine, heavier springs and shock absorbers, a dual exhaust and low-profile tires. All this was astonishing value for the extra $295 that Pontiac was asking. The production series GTO could cover the sacrosanct 'standing quarter' at the dragstrip in 15 seconds, a full 5 seconds faster than a Ferrari 275GTB!

This obvious winner in the 'Stop-light Grand Prix' was an instant success. Sales rocketed right from the start and the Pontiac GTO became the first in a long line of American musclecars.

The Pontiac GTO seemed to have a license to fly: it could cover the standing quarter mile faster than a Ferrari 275GTB. Its designers were trying to cater for teenagers who were just learning to drive. They had come up with the solution of putting a V8 engine in a compact body with special suspension and tires, and a body trimmed back to essentials. The result was a roaring success.

GTO

PORSCHE

Porsche Speedster

The Porsche Speedster is a very strange-looking machine. It's as if the roof of a 356 coupe has been sliced off with a can-opener, leaving an open cockpit which is about as luxuriously appointed as a kamikaze bomber. The windshield isn't much better – it looks as if the Zuffenhausen factory installed a Goggomobile version instead. Yet this curious combination exudes a perverse charm that is very hard to resist. Indeed, no other model of Porsche has contributed as much as the Speedster to the enduring Porsche mystique.

At the beginning of the 1950s, the success of British sportscars in the United States market made other car manufacturers want a slice of the cake. Very quickly, a whole host of sportscars appeared, trying, with varying degrees of success, to correspond to the European idea of what the American idea of a sportscar should be. Rereading that sentence makes it clear that the results were likely to be questionable.

Porsche presented its Speedster at the Brussels Motor Show early in 1955. Like the Mercedes 300SL, the Lancia Aurelia Spyder or the BMW 507, the Speedster had been the brainchild of Porsche's agent in New York, Max Hoffman, the 'Baron of Park Avenue.' For its part, head office had reasoned along these lines: Porsches were such special cars that their future drivers had to be wooed at a very tender age, before they had had time to develop bad driving habits at the wheel of an ordinary car.

In order to capture the youth market the Speedster would need to have two things going for it: attractive styling and a price tag under $3000. The low price was achieved by reducing the car's equipment to essentials: removable rather than wind-down windows and an umbrella-like fold-down hood (which also represented a considerable saving in weight). Whether the goal of attractive styling was met was open to question. Quite clearly, the men at the factory in Zuffenhausen, Germany did not feel comfortable about constructing a toy car, but in the end this only added to the car's crackpot charm. With its not-quite-big-enough windshield and its strange roof, the Speedster seemed quite unique.

With a special twin overhead cam and roller-crankshaft, the 110hp engine could whiz the very light Speedster along at speeds of up to 120mph. Whether new or secondhand, the Speedster was the best possible introduction, for thousands of young Americans, to the fascinating experience of driving one of Doctor Porsche's automobiles.

Porsche offered the buyer four versions of its Speedster: the 1500 Standard (55hp), the 1500 Super (70hp), the 1600 Standard (60hp) or the 1600 Super (75hp). The car shown here is the 1600 Super. The version that contributed most to the Porsche mystique was the 1500 Super equipped with the famous Hirth roller-bearing crankshaft. This was reputed to selfdestruct anywhere above 2500rpm because of an unfortunate vibration period. This quirk of behavior obviously encouraged sporty driving! By paying extra (in fact, almost double . . .) your Speedster could be delivered with a Carrera-type engine (1500 or 1600 version) with twin overhead cam head and Hirth crankshaft. These engines developed 100 and 110hp respectively and the car could reach 120mph.

Inevitably, the Speedster displayed the exuberant tendency to oversteer that was a common failing of the early Porsches. It was not a car for Sunday drivers . . . But then everybody knew that. And the proud owner behind the wheel of his Speedster knew that everybody knew it – a little extra source of satisfaction. From late 1954 until the beginning of 1959, about 5000 Porsche Speedsters were built. Yet the Speedster did much more for Porsche's reputation than for the company's balancesheets – the low price left a very meager profit margin for the maker. Porsche later decided to recoup on the financial side by marketing a more luxuriously equipped convertible that could justifiably be given a much higher price tag.

Porsche 356

In 1946, Porsche's research and design team was to be found in the small town of Gmund in the Austrian Alps, without offices, without staff and without funds. Then Ferry Porsche, the professor's son, heard of the firm of Cisitalia. The Italians had made a successful racing debut with cars built from Fiat components. Now the firm's founder, Piero Dusio, wanted to try his luck at Formula 1 and was looking for a design team to develop the car he needed. Ferry Porsche came up with a project of quite bewildering complexity: 12 cylinders horizontally-opposed, two superchargers and 4-wheel-drive! The stunned Italian agreed. From that point on, the Italian firm started to decline while Porsche went from strength to strength. The cost of the project landed up at four or five times the original estimate, but the monster single-seater never saw the light of day. In the end, the Italians went bankrupt and Porsche could at last turn its attention to the realization of its own projects.

The association with Cisitalia had been beneficial to Ferry Porsche in a totally different way too – he had been able to observe how the Italians built excellent sportscars from standard Fiat components. Porsche decided to follow their example. There were no Fiats in Germany, but there were plenty of Volkswagens . . . So the design team started on Project 356.

The first Porsche 356 was completed by May 1948. It had a tubular chassis like the Cisitalia cars and a Volkswagen engine and transmission, but the power unit was reversed to put the engine forward of the wheels. This car was soon followed by a coupe which, with its overhanging rear engine and platform chassis, was much more like a Volkswagen. The body had been styled by Erwin Kommenda and was inspired by Professor Porsche's project for a racing car based on the Volkswagen. This was the car that was to be modified and improved every year from then on until 1965, to finally become the 95hp but very light 1600SC, capable of a top speed of 110mph.

The Porsche 356SC released in 1963 was the definitive version of the original 356 that appeared in 1948. It was still being produced until the end of 1965 alongside the new 911. The engine from the 356 lived on in the 912, which was a 4-cylinder version of the 911. In 1969, the 912 gave way to the 914 and the link with the Volkswagen at last seemed to have been broken. But not quite: the 914/4 had a Volkswagen engine. The 1600SC engine developed 95hp and enabled the little car to reach a top speed of 110mph. This model had Ate disc brakes on all four wheels, with the characteristic drums added at the rear for the handbrake. The car shown here has a modified exhaust and bumper guards: on the original models, the exhaust pipes stick right through the bumper guards.

Porsche 911 Turbo

The 911 Turbo shown here is a 1982 model. At $45,000 it cost twice as much as a 911SC coupe at a time when the Lamborghini Countach cost $80,000. Five years later, all these price tags had easily doubled. Yet since the 911 Turbo was released, sales have remained fairly stable, between 1200 and 1500 per year. The Turbo's top speed of 160mph is one of the most impressive offered in the automobile industry. Like Ford with its Model T or Volkswagen with its Beetle, Porsche finds itself a prisoner of the continuing success of the 911 – neither the 928 nor its S version has managed to eclipse the 911, which in 1985 gave birth to a cabriolet (see previous pages).

In a contemporary production of *Cinderella* the pumpkin could be represented by an old and battered Volkswagen Beetle, and Cinderella's gorgeous coach by a Lamborghini Countach. Yet it is hardly an exaggeration to say that the driver of a 911 Turbo can take part in a fairytale adventure too. The magic begins with the accelerator pedal. Below 3000rpm, the Turbo is no more than an ordinary sportscar that sounds just a little tinny. Around 3500rpm the Turbo passes into a world of enchantment. The car takes on the power of a jet and the metallic sound of its exhaust develops into a throaty roar as the increasing volume of exhaust gases speeds up the turbocharger.

For the men at Porsche, creation is a continuing process rather than something based simply on a flash of inspiration. It is very difficult to establish the genealogy of one model without taking into account all the earlier models. This is no easy task, especially given the complex relationships between the different versions of the 911. The easiest way to define the 911 Turbo would be to call it a turbo version of the 3-liter Carrera, of which 159 were built for the 1974 racing season. The Porsche factory also built (not for sale) some 2.1-liter Carreras with turbochargers, where the turbo experience acquired with the 917 was applied for the first time to a horizontally-opposed 6-cylinder engine. The results were encouraging enough for Porsche to try the technique on a 3-liter engine. That car was christened the 911 Turbo and went on display at the Paris Motor Show in 1974.

The display car was not much more than a camouflaged 3-liter Carrera. All that Porsche intended to do was build 400 of them, lightened and simplified rather like the Carrera. Then Ernst Fuhrmann, who was at the head of Porsche at the time, decided that these cars were to be fully and luxuriously equipped, with electric windows, air-conditioning, fully-padded interior and thick carpets. A nearly definitive version of this 911 Turbo went on show at the Geneva Motor Show in March 1975. The one feature which was not definitive was the gearbox. The Carrera box was not quite up to handling the torque of the turbo engine; the car needed a completely redesigned 4-speed gearbox. The 911 Turbo also had to wait until the following year for tires that were worthy of it – the famous Pirelli P7s. The brakes were finally perfected on the 1978 model, which was also given an engine increased to 3.3 liters, capable of pushing a top speed of 160mph.

The 911 Turbo was the first of Professor Porsche's line to be admitted to that very exclusive club reserved for 'supercars.'

9693 XC 95

Porsche 928

Having shown with the 911 Turbo that a Volkswagen could reach speeds of more than 155mph, Porsche decided that they had made their point. Their next car, released in 1977, would have to represent the state of the art.

Indeed, the 928 had a personality as slick as its bodystyling. It was a streamlined, air-conditioned, stereophonic car designed to speed effortlessly along the fast lane of a freeway – it was smooth, silent and very efficient.

Up front, the 928 had a compact water-cooled V8 linked, like that of the 924, by a fixed shaft to the axle gearbox mounted at the rear. Porsche was proclaiming to his customers that he had created the almost-perfect automobile, better even than the 911. The 928 order book was open. Porsche was quite right – the 928 was indeed the almost-perfect car. By comparison, the 911 was restive and unruly. So one can understand the consternation at Zuffenhausen when, after politely admiring the 928, the faithful continued buying the 911 just as before! Porsche took the point and added a few sharp edges to the 928's bodystyling and personality. The result was the 928S released in 1980, a 300hp machine that could whisk you along at a comfortable 155mph. Everything that the 911 could do in an atmosphere of furious excitement, the 928S could do with quiet assurance.

The Porsche 928, which was released in March 1977, was to be the third generation Porsche, the one that would finally exorcize the ghost of Ferdinand Porsche and his rear-mounted, air-cooled engine. But Porsche's customers decided otherwise, and the 911 retained its faithful following. The 928 still managed to carve out for itself a slice of the luxury GT market, but it did so at the expense of other German cars rather than the more traditional Porsches. In 1980, the 240hp 928 was joined by the 928S with its increased power-rating of 300hp. The model shown here is the 928; the only other version to be built was the 928S.

The 928 shown here was followed by the 928S. From the outside, the 928S can be distinguished from the previous 928 by its discreet aerodynamic accessories – the air dam and rear spoiler, which add to the car's appeal and also encourage stability at high speed. With a top speed in excess of 155mph, the 928S eats up miles on the freeway with enormous ease and style.

4840GA75

THUNDERBIRD

Ford Thunderbird

The 2-seater version (1955 to 1957) of the Thunderbird created the legend. Its 5.1-liter engine developed 215hp and could launch the car from 0 to 60mph in 11 seconds. After that, the designers rested on their laurels. This early success spoiled the later Thunderbirds: compare the 1956 model on the previous pages with the 1963 T-bird shown here. The first displays graceful youth, the second gross maturity. The 1963 model was also available in a roadster version.

In 1951, a somewhat intimidating vice-president of the Ford Corporation was visiting the Paris Motor Show with one of his colleagues. When he caught sight of the Jaguar XK 120 he stood wide-eyed with amazement and exclaimed angrily, 'Why doesn't Ford have a car like that?' He turned around, but his fearful associate had vanished. A few minutes later, he returned to reassure his chief that Ford did, indeed, have a sportscar currently on the drawing board. 'Currently' really meant 'as of two minutes ago' – he had just made a long-distance phonecall to Dearborn to set the Thunderbird project in motion.

After getting off the mark with much squealing of tires, the project quickly lost momentum. In the early 1950s, Ford had other projects on the go. The Corporation was losing a lot of sleep over its rival Chevrolet. Detroit just wasn't big enough for both of them, and Ford was ready to act at the slightest provocation.

Then in 1953 Chevrolet unveiled the Corvette. Dearborn immediately set about putting the sportscar project into action. First of all they needed a name. It was a young designer holidaying near the Mexican border who came up with the mythical 'thunderbird' which was used to decorate motels and hotdog stands in the area. The car itself would simply be a shortened, lowered, open version of the 1955 range of Ford saloons. It just so happened that that year's Fords were particularly attractive, with straight fins, round taillamps and a pseudo-Ferrari radiator grille. At the same time the firm had just put the finishing touches to a new rocker-arm V8 that, compared to the old one, was a real advance on the existing models. The Thunderbird was to be equipped with the most powerful 5.1-liter, 215hp version, the one in the Mercury. Styling for the Thunderbird was left to the Ford designers working under Frank Hershey – the result was a big car, 15 feet in length. The project turned out to be brilliantly successful, except that the 110mph Thunderbird wasn't really a sportscar at all.

0 20 40 60 80 100 120 140 160 180 200
OIL
TEMP
E FULL
LIGHTS
WIPERS WASHERS
LEFT AIR RIGHT AIR

GA 75 F